TEACHER GUIDE

Includes St
Workshe

MW00335107

7th–8th Grade

Science

Quizzes

Introduction to Astronomy

MASTER BOOKS
— CURRICULUM —

Author: Jason Lisle

Master Books Creative Team:

Editor: Craig Froman

Design: Terry White

Cover Design: Diana Bogardus

Copy Editors:
Judy Lewis
Willow Meek

Curriculum Review:
Kristen Pratt
Laura Welch
Diana Bogardus

First printing: August 2016
Sixth printing: April 2022

Master Books® is a division of the New Leaf Publishing Group, Inc.

ISBN: 978-0-89051-990-5
ISBN: 978-1-61458-569-5 (digital)

Printed in the United States of America

Please visit our website for other great titles:
www.masterbooks.com

About the Author

Dr. Jason Lisle is a Christian astrophysicist who writes and speaks on various topics relating to science and the defense of the Christian faith. He graduated from Ohio Wesleyan University where he majored in physics and astronomy and minored in mathematics. He then earned a master's degree and a Ph.D. in astrophysics at the University of Colorado in Boulder. Dr. Lisle began working in full-time apologetics ministry, specializing in the defense of Genesis. His most well-known book, *The Ultimate Proof of Creation*, demonstrates that biblical creation is the only logical possibility for origins.

I'm loving this whole line so much. It's changed our homeschool for the better!

—Amy ★★★★★

Your reputation as a publisher is stellar. It is a blessing knowing anything I purchase from you is going to be worth every penny!

—Cheri ★★★★★

Last year we found Master Books and it has made a HUGE difference.

—Melanie ★★★★★

We love Master Books and the way it's set up for easy planning!

—Melissa ★★★★★

You have done a great job. MASTER BOOKS ROCKS!

—Stephanie ★★★★★

Physically high-quality, Biblically faithful, and well-written.

—Danika ★★★★★

Best books ever. Their illustrations are captivating and content amazing!

—Kathy ★★★★★

Affordable
Flexible
Faith Building

MASTERBOOKS
— CURRICULUM —

Table of Contents

Features: The suggested weekly schedule enclosed has easy-to-manage lessons that guide the reading, worksheets, and all assessments. The pages of this guide are perforated and three-hole punched so materials are easy to tear out, hand out, grade, and store. Teachers are encouraged to adjust the schedule and materials needed in order to best work within their unique educational program.

All About Astronomy! Have you ever wondered which stars are closest to Earth? Or which constellations you can view in different seasons? What about the best telescopes to use for stargazing? Find the answers to these questions and more in this intriguing course about our night sky! From planets to nebulas, this course will teach students how to find and identify various celestial objects, using telescopes, binoculars, or even just the naked eye. The full-color star charts and the Stargazer's Planisphere included in *The Stargazer's Guide to the Night Sky* help students locate star positions on any night of the year.

🕐	Approximately 30 to 45 minutes per lesson, two to three days a week
🔑	Includes answer keys for worksheets and quizzes
📝	Worksheets for each chapter
🔁	Quizzes are included to help reinforce learning and provide assessment opportunities
📄	Designed for grades 7 to 8 in a one-year science course
⚗	Suggested labs (if applicable)

Course Objectives: Students completing this course will

- Evaluate how the phases of the moon work
- Discover how to choose the best telescope for them
- Investigate the motions of the sky, star classification, and deep sky objects

- Identify what sort of objects can be seen with binoculars
- Learn the best ways and optimal times to observe planets and stars

Course Description

Explore the night sky: identify stars, constellations, and even planets. Learn how to stargaze with a telescope, binoculars, or even your naked eye. Allow Dr. Jason Lisle, a research scientist with a master's and PhD in astrophysics, to guide you in examining the beauty of God's Creation. Find Orion, a well-known constellation; within it you can easily see the red super giant star Betelgeuse over 3,000 trillion miles away without binoculars or a telescope! At 60,000 times the diameter of Earth, it is a celestial sight! Or marvel at our galaxy, the Milky Way, shining brightly overhead in late summer as you see the dark patches of dust as well as light spots containing entire star clusters! *The Stargazer's Guide to the Night Sky* includes 150 beautiful, full-color star charts and other easy-to-use illustrations for success. When will the next solar eclipse take place? What is that bright star setting in the west? How do I find Saturn? Take a few moments to stand and look up at the glorious night sky, appreciating the majestic beauty of God's vast universe. The student book also includes the Stargazer's Planisphere, a chart that helps you locate the positions of stars on any night of the year so you can better enjoy God's amazing night sky.

Course Credit

Parents may assign this either as a Science or Elective credit. The description "with Lab" may be added to course title if student completes about 15 hours of lab work. While labs are not included in this course, it can easily be supplemented with labs if the teacher would like to do so. A variety of astronomy labs are available online, and students can also use informative apps for their phone or tablet for additional assistance in finding objects in the night sky. Optional Course Equipment: A good telescope is highly recommended.

First Semester Suggested Daily Schedule

Date	Day	Assignment	Due Date	✓	Grade
		First Semester–First Quarter			
Week 1	Day 1	Read Pages 4–5 • *The Stargazer's Guide to the Night Sky* • (SGNS)			
	Day 2				
	Day 3	Intoduction Stargazer Introduction: Worksheet 1 • Pages 15–16 Teacher Guide • (TG)			
	Day 4				
	Day 5	Read Pages 6–10 • (SGNS)			
Week 2	Day 6				
	Day 7	Motions in the Sky — Basic Stargazer Ch1: Worksheet 1 • Pages 17–18 • (TG)			
	Day 8				
	Day 9	Read Pages 11–15 • (SGNS)			
	Day 10				
Week 3	Day 11	Motions in the Sky — Basic Stargazer Ch1: Worksheet 2 • Pages 19–20 • (TG)			
	Day 12				
	Day 13	Read Pages 16–20 • (SGNS)			
	Day 14				
	Day 15	Motions in the Sky — Basic Stargazer Ch1: Worksheet 3 • Pages 21–22 • (TG)			
Week 4	Day 16				
	Day 17	Read Pages 21–23 • (SGNS)			
	Day 18				
	Day 19	Motions in the Sky — Basic Stargazer Ch1: Worksheet 4 • Pages 23–24 • (TG)			
	Day 20				
Week 5	Day 21	Read Pages 24–27 • (SGNS)			
	Day 22				
	Day 23	Motions in the Sky — Advanced Stargazer Ch2: Worksheet 1 • Pages 25–26 • (TG)			
	Day 24				
	Day 25	Read Pages 28–33 • (SGNS)			
Week 6	Day 26				
	Day 27	Motions in the Sky — Advanced Stargazer Ch2: Worksheet 2 • Pages 27–28 • (TG)			
	Day 28				
	Day 29	Read Pages 34–37 • (SGNS)			
	Day 30				

Date	Day	Assignment	Due Date	✓	Grade
Week 7	Day 31	Motions in the Sky — Advanced Stargazer Ch2: Worksheet 3 • Page 29 • (TG)			
	Day 32				
	Day 33	Read Pages 38–41 • (SGNS)			
	Day 34				
	Day 35	Motions in the Sky — Advanced Stargazer Ch2: Worksheet 4 • Page 31 • (TG)			
Week 8	Day 36				
	Day 37	*The Stargazer's Guide to the Night Sky*: **Chapters 1–2 Quiz** Pages 101–104 • (TG)			
	Day 38				
	Day 39	Read Pages 42–47 • (SGNS)			
	Day 40				
Week 9	Day 41	Understanding the Eye Stargazer Ch3: Worksheet 1 • Page 33 • (TG)			
	Day 42				
	Day 43	Read Pages 48–53 • (SGNS)			
	Day 44				
	Day 45	Astronomy with the Unaided Eye Stargazer Ch4: Worksheet 1 • Page 35 • (TG)			
First Semester–Second Quarter					
Week 1	Day 46	Read Pages 54–61 • (SGNS)			
	Day 47				
	Day 48	Astronomy with the Unaided Eye Stargazer Ch4: Worksheet 2 • Page 37 • (TG)			
	Day 49				
	Day 50	Read Pages 62–69 • (SGNS)			
Week 2	Day 51				
	Day 52	Astronomy with the Unaided Eye Stargazer Ch4: Worksheet 3 • Page 39 • (TG)			
	Day 53				
	Day 54	*The Stargazer's Guide to the Night Sky*: **Chapters 3–4 Quiz** Pages 105–106 • (TG)			
	Day 55				
Week 3	Day 56				
	Day 57	Read Pages 70–74 • (SGNS)			
	Day 58				
	Day 59	Celestial Events Stargazer Ch5: Worksheet 1 • Page 41 • (TG)			
	Day 60				

Date	Day	Assignment	Due Date	✓	Grade
Week 4	Day 61	Read Pages 75–79 • (SGNS)			
	Day 62				
	Day 63	Celestial Events Stargazer Ch5: Worksheet 2 • Page 43 • (TG)			
	Day 64				
	Day 65	Read Pages 80–83 • (SGNS)			
Week 5	Day 66				
	Day 67	Celestial Events Stargazer Ch5: Worksheet 3 • Page 45 • (TG)			
	Day 68				
	Day 69	Read Pages 84–87 • (SGNS)			
	Day 70				
Week 6	Day 71	Celestial Events Stargazer Ch5: Worksheet 4 • Page 47 • (TG)			
	Day 72				
	Day 73	Read Pages 88–91 • (SGNS)			
	Day 74				
	Day 75	Telescope Basics Stargazer Ch6: Worksheet 1 • Page 49 • (TG)			
Week 7	Day 76				
	Day 77	Read Pages 92–96 • (SGNS)			
	Day 78				
	Day 79	Telescope Basics Stargazer Ch6: Worksheet 2 • Page 51 • (TG)			
	Day 80				
Week 8	Day 81	Read Pages 97–100 • (SGNS)			
	Day 82				
	Day 83	Telescope Basics Stargazer Ch6: Worksheet 3 • Page 53 • (TG)			
	Day 84				
	Day 85	Read Pages 101–105 • (SGNS)			
Week 9	Day 86				
	Day 87	Telescope Basics Stargazer Ch6: Worksheet 4 • Page 55 • (TG)			
	Day 88				
	Day 89	*The Stargazer's Guide to the Night Sky*: **Chapters 5–6 Quiz** Pages 107–108 • (TG)			
	Day 90				
		Mid-Term Grade			

Second Semester Suggested Daily Schedule

Date	Day	Assignment	Due Date	✓	Grade
		Second Semester–Third Quarter			
Week 1	Day 91				
	Day 92	Read Pages 106–108 • (SGNS)			
	Day 93				
	Day 94	Telescope Observing Sessions Stargazer Ch7: Worksheet 1 • Page 57 • (TG)			
	Day 95				
Week 2	Day 96	Read Pages 109-115 • (SGNS)			
	Day 97				
	Day 98	Telescope Observing Sessions Stargazer Ch7: Worksheet 2 • Page 59 • (TG)			
	Day 99				
	Day 100	Read Pages 116–121 • (SGNS)			
Week 3	Day 101				
	Day 102	Telescope Observing Sessions Stargazer Ch7: Worksheet 3 • Page 61 • (TG)			
	Day 103				
	Day 104	Read Pages 122–127 • (SGNS)			
	Day 105				
Week 4	Day 106	The Moon and the Sun Stargazer Ch8: Worksheet 1 • Page 63 • (TG)			
	Day 107				
	Day 108	Read Pages 128–133 • (SGNS)			
	Day 109				
	Day 110	The Moon and the Sun Stargazer Ch8: Worksheet 2 • Page 65 • (TG)			
Week 5	Day 111				
	Day 112	*The Stargazer's Guide to the Night Sky*: **Chapters 7–8 Quiz** Page 109 • (TG)			
	Day 113				
	Day 114	Read Pages 134–138 • (SGNS)			
	Day 115				
Week 6	Day 116	The Planets Stargazer Ch9: Worksheet 1 • Page 67 • (TG)			
	Day 117				
	Day 118	Read Pages 139–143 • (SGNS)			
	Day 119				
	Day 120	The Planets Stargazer Ch9: Worksheet 2 • Page 69 • (TG)			

Date	Day	Assignment	Due Date	✓	Grade
	Day 121	Read Pages 144–148 • (SGNS)			
	Day 122				
Week 7	Day 123	The Planets Stargazer Ch9: Worksheet 3 • Page 71 • (TG)			
	Day 124				
	Day 125	Read Pages 149–152 • (SGNS)			
	Day 126				
	Day 127	The Planets Stargazer Ch9: Worksheet 4 • Page 73 • (TG)			
Week 8	Day 128				
	Day 129	Read Pages 153–157 • (SGNS)			
	Day 130				
	Day 131	The Planets Stargazer Ch9: Worksheet 5 • Page 75 • (TG)			
	Day 132				
Week 9	Day 133	Read Pages 158–163 • (SGNS)			
	Day 134				
	Day 135	The Planets Stargazer Ch9: Worksheet 6 • Page 77 • (TG)			
Second Semester–Fourth Quarter					
	Day 136	Read Pages 164–167 • (SGNS)			
	Day 137				
Week 1	Day 138	Star Classifications and Telescope Viewing Stargazer Ch10: Worksheet 1 • Page 79 • (TG)			
	Day 139				
	Day 140	Read Pages 168–175 • (SGNS)			
	Day 141				
	Day 142	Star Classifications and Telescope Viewing Stargazer Ch10: Worksheet 2 • Page 81 • (TG)			
Week 2	Day 143				
	Day 144	*The Stargazer's Guide to the Night Sky*: Chapters 9–10 Quiz Pages 111–112 • (TG)			
	Day 145				
	Day 146				
	Day 147	Read Pages 176–189 • (SGNS)			
Week 3	Day 148				
	Day 149	Deep Sky Objects Stargazer Ch11: Worksheet 1 • Page 83 • (TG)			
	Day 150				

Date	Day	Assignment	Due Date	✓	Grade
	Day 151	Read Pages 190–196 • (SGNS)			
	Day 152				
Week 4	Day 153	Deep Sky Objects Stargazer Ch11: Worksheet 2 • Page 85 • (TG)			
	Day 154				
	Day 155	Read Pages 197–203 • (SGNS)			
	Day 156				
	Day 157	Deep Sky Objects Stargazer Ch11: Worksheet 3 • Page 87 • (TG)			
Week 5	Day 158				
	Day 159	Read Pages 204-211 • (SGNS)			
	Day 160				
	Day 161	Deep Sky Objects Stargazer Ch11: Worksheet 4 • Page 89 • (TG)			
	Day 162				
Week 6	Day 163	Read Pages 212–221 • (SGNS)			
	Day 164				
	Day 165	Deep Sky Objects Stargazer Ch11: Worksheet 5 • Page 91 • (TG)			
	Day 166				
	Day 167	Read Pages 222–227 • (SGNS)			
Week 7	Day 168				
	Day 169	Astrophotography Stargazer Ch12: Worksheet 1 • Page 93 • (TG)			
	Day 170				
	Day 171	Read Pages 228–233 • (SGNS)			
	Day 172				
Week 8	Day 173	Astrophotography Stargazer Ch12: Worksheet 2 • Page 95 • (TG)			
	Day 174				
	Day 175	Read Pages 234–235 • (SGNS)			
	Day 176				
	Day 177	The Relevance of Astronomy Stargazer Ch12: Worksheet 3 • Page 97 • (TG)			
Week 9	Day 178				
	Day 179	*The Stargazer's Guide to the Night Sky*: **Chapters 11–12 Quiz** • Page 113 • (TG)			
	Day 180				
		Final Grade			

Astronomy Worksheets

for Use with

The Stargazer's Guide to the Night Sky

Short Answers

1. What are you expecting and hoping to learn from this course?

2. What are three different ways you can observe the night sky?

3. Is there a difference in the sky depending on whether you live in the Southern or Northern Hemisphere?

4. Is there a difference in the sky depending on the season?

5. What are the two largest celestial objects we can view?

Short Answers

1. Why is there a trend of east-to-west motion when observing the sun, moon, and stars?

2. What is this trend of the earth's rotation called?

3. What is the approximate rotation time of the earth that relates to other objects in the sky as well?

4. Name the concept that is useful for understanding the positions and motions of stars.

5. Name the concept that involves expanding the earth's equator into space.

6. What are constellations called that are close enough to the celestial pole that they are visible all night, year-round?

7. Since the 23 hours and 56 minutes period is the length of time it takes the earth to turn as seen from a distant star, this is called a "_____ _____."

Short Answers

1. Describe the difference between the sidereal day and the solar day.

2. Since stars rise two hours earlier every month, they rise ____ hours earlier after six months.

3. Why is it difficult to observe planets, globular clusters, and nebulae when the moon is full?

4. The moon rises (on average) about _____ _____ later each day.

5. Why is it ironic that the moon is called "the moon" in regards to the gravitational pull of the earth and sun?

6. The phases of the moon are not related to the earth's shadow, but to the _____ of the day side of the moon we can see from our position.

7. It takes _____ days for the moon to go through its phases, and _____ days for its orbital period.

Short Answers

1. The motion of the planets is complicated because their apparent motion in the sky is the combination of their _____ _____ around the sun, plus the _____ _____ in position due to Earth's motion around the sun.

2. What is responsible for the seasons on Earth?

3. Describe what the two coordinate systems widely used in astronomy are based on.

4. _____ describes how high above the horizon an object is (in angle).

5. _____ describes how far along the horizon an object is to the right of due north.

6. Equatorial coordinates are based on the _____ _____. In particular, they are based on the celestial _____.

Short Answers

1. Stars with a declination that is less than your latitude will pass _____ of zenith when they cross the meridian; stars with a greater declination will pass _____ of zenith.

2. Which RA coordinates can be seen depends on the time of _____ and the time of _____.

3. Once calibrated on an object whose RA and Dec you know, you can use _____ circles on a telescope to find any other RA or Dec.

4. What is the best way to get a feel for the motions in the sky?

5. A star wheel or planisphere helps you find stars by lining up the _____ with the _____ on the planisphere.

6. A planisphere helps find constellations, but will not help with _____ or the _____.

Short Answers

1. The sun's _____ is zero on the vernal equinox or autumnal equinox, so the sun is above the horizon for exactly _____ hours.

2. What does the term "equinox" mean?

3. What does the term "solstice" mean?

4. Anyone observing from the _____ Circle will experience 24 hours of sunlight on the summer solstice.

5. In the Northern Hemisphere, days are longer in the _____ and _____ seasons, and shorter in the _____ and _____ seasons.

6. The moon orbits roughly in the _____, not the celestial equator.

Short Answers

1. When will the first quarter moon make its highest arc in the sky?

2. There are two kinds of eclipses: (1) A _____ eclipse, when the moon's shadow falls on the earth, and (2) a _____ eclipse, when the earth's shadow falls upon the moon.

3. The term "_____" is the generic term for the two points of intersection on any two great circles.

4. The configuration of the sun, moon, and Earth is about the same every 18.031 years, and is called the _____ cycle.

5. The earth actually has two shadows. A darker inner shadow called the "_____" and a lighter outer shadow called the "_____."

6. During a total solar eclipse, the bright surface of the sun (the _____) is completely blocked by the moon.

7. Any given spot on the earth experiences a total solar eclipse once every _____ years or so.

Short Answers

1. When the moon is farther from the earth, it appears smaller than the sun, and when it passes directly in front of the sun it leaves a thin "ring" or "_____."

2. The moon seems to wobble a bit from week to week, which is called "_____."

3. Planets orbit the sun in slightly _____ paths with the sun at one focus of the ellipse.

4. The _____ planets are Mars, Jupiter, Saturn, Uranus, and Neptune.

5. _____ is when a planet is "behind" the sun.

6. All the planets orbit the sun in the same direction: _____.

7. _____ and _____ are the two inferior planets whose orbits lie closer to the sun than Earth does.

Short Answers

1. A _____ transit is when Mercury or Venus cross directly in front of the sun during inferior conjunction.

2. Comets are icy objects and generally have extremely _____ orbits.

3. Meteor showers are generally caused by debris left behind by a _____.

4. The most impressive, reliable meteor shower is the _____ meteor shower, occurring around August 12th each year.

5. The sun traces out a thin figure-eight shape across the sky with each day, which is called_____.

6. The constellation in which the sun is found at equinox shifts with time. This phenomenon is called the "_____ _____ _____ _____."

Short Answers

1. The human eye has two different types of light-detecting cells: _____ and _____.

2. Rods are far more sensitive to _____ than cones, and are thus able to detect much fainter objects.

3. When looking at faint objects through a telescope, you must avoid looking _____ at the object.

4. The rods contain a chemical (_____) that is light sensitive.

5. Using a _____ _____ at night helps you see without impacting your dark-adapted vision.

6. As a general guideline, it takes about ____ minutes to dark adapt your eyes.

7. Eating _____ really does help improve your night vision because of the carotene in them.

Short Answers

1. Stars are named based on their _____ and the _____ in which they are found.

2. The constellation Orion is mentioned in the biblical book of _____, dating back to about 2000 B.C.

3. _____ are groups of bright stars that are not one of the official 88 constellations.

4. Mintaka is a blue bright giant, so one of the _____, _____ stars in the night sky.

5. Because its light fades and grows, Algol is often called the "_____ Star."

6. _____ is the brightest star visible in Earth's nighttime sky.

7. Since it is in the constellation Canis Major, Sirius is sometimes referred to as the "_____ _____."

Short Answers

1. In mid-April, the constellation that dominates the evening sky is _____.

2. Polaris or the "_____ _____" is the one star that doesn't move noticeably with the time of night or time of year.

3. _____ is the center star in the handle of the Dipper.

4. _____ used to be the North Star around the year 2800 B.C.

5. The constellation _____ is dominated by the bright red supergiant Antares.

Short Answers

1. The Summer _____ is made up of Vega, Altair, and Deneb.

2. The brightest star in the constellation Cygnus is _____, which means the "tail."

3. The constellation Corona Australis is also known as the _____ _____.

4. The Great _____ comprises Markab, Scheat, Algenib, and Alpheratz.

5. The constellation Andromeda is shaped like a curved _____.

6. At a distance of 2.9 million light years, _____, or "the Andromeda Galaxy," is the farthest and largest object that can easily be seen with the unaided eye.

Short Answers

1. A _____ is when one celestial body passes closely by another.

2. A _____ occurs when three or more objects come close together in the sky.

3. A _____ _____ occurs when a planet passes by another celestial object three times in a row.

4. An _____ is when one celestial object passes directly in front of another, blocking the background object from sight.

5. Only _____ eclipses are true eclipses, where we observe the shadow of one body falling on another.

6. Total _____ eclipses are far more common than total _____ eclipses.

Short Answers

1. The _____ method is a way to safely view the sun during a solar eclipse.

2. A meteor, sometimes called a "falling star," is caused by a small bit of matter falling from space and burning up as it is _____ by atmospheric resistance.

3. The term "_____" refers to the bright streak in the sky. The rock causing the streak is called the "_____." The rock that hits the earth's surface is called a "_____."

4. During a meteor shower, meteors will appear to move away from the same point in space. This point is called the "_____."

5. A particularly bright meteor is called a "_____."

6. The visible trail that glows behind a bright meteor is called a "_____."

7. A "_____ _____" is a rare occurrence where the sky appears to be raining meteors.

Short Answers

1. The only meteor shower known to be associated with an asteroid rather than a comet is the
   _____.

2. The _____ are the best winter meteor showers.

3. The two meteor showers associated with Halley's Comet are the _____ _____ and the
   _____.

4. Approximately every 33 years, the _____ produce a meteor storm.

5. The reason rainbows are colorful is because the water droplets split the sunlight into its constituent
   _____.

6. A _____ _____ is caused by the reflection and refraction of sunlight off of ice crystals in
   thin, high cirrus clouds.

Short Answers

1. The word *corona* means "_____."

2. Both _____ _____and _____ are caused by sunlight reflecting off of dust in the solar system.

3. _____ orbit the earth in about 90 minutes.

4. _____ is the worst time to try to view satellites.

5. The iridium _____ is the temporary brightening of a satellite associated with the company Iridium Communications Inc.

Short Answers

1. One of the telescope's most important functions is to make faint things _____ enough to be seen.

2. For the telescope, _____ power is the ability to "separate" things that are close together.

3. The overall ability of a telescope is determined by the diameter of the primary lens or primary mirror — the "_____."

4. The two kinds of telescopes are _____ and reflectors.

5. _____ use curved mirrors to reflect and focus incoming light.

6. _____ use lenses only — no mirrors.

7. "_____ aberration" is when bright objects you view through a refractor are surrounded by a little purple halo.

Short Answers

1. Reflectors use a curved _____ instead of a primary lens to bring light to a focus.

2. Newtonians are reflecting telescopes named after _____ _____.

3. The _____-_____ is generally considered to be the best all-around general-use telescope for amateur astronomers.

4. "_____" is when dew collects on the corrector plate of a telescope.

5. A "_____ _____" can prevent dew from collecting as you observe the night sky.

Short Answers

1. Binoculars or a small telescope will allow you to observe the moon, bright stars, open star clusters, the _____ Galaxy, and some bright comets.

2. A _____ allows the telescope to pivot in at least two directions.

3. A _____-_____ is a motor that slowly rotates the telescope in the opposite direction that Earth rotates.

4. Telescope mounts come in two varieties: _____ mounts and _____ (alt-azimuth) mounts.

5. A _____ telescope is the name of the second small refractor telescope often attached to the larger one.

6. A _____ adds a laser-produced artificial spot, concentric circles, or crosshairs that indicate exactly where the telescope is pointed.

7. If you want to do astrophotography, get a _____-controlled system.

Short Answers

1. The most important tip for setting up a telescope for the first time is to set it up _____ _____ _____.

2. You can begin the alignment of the telescope as soon as it is dark enough to see the _____ _____.

3. If a telescope is computer controlled, turn it on, and begin the _____-_____ program.

4. Spotter scopes (1)_____, (2)_____, and (3)_____ the image.

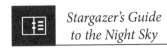
Short Answers

1. The clearest nights will generally be the _____ nights.

2. A _____ flashlight is best because the colored light will not spoil dark adaptation.

3. A _____ laser pointer is useful for pointing out celestial objects to other people.

4. If the telescope temperature does not match the air temperature, all objects will appear fuzzy and "wavy" due to _____ within the telescope.

5. Expect about a _____ _____ to become more or less fully dark adapted.

6. A _____ _____ drastically reduces the amount of heat the corrector plate can radiate to space.

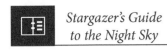
Short Answers

1. _____ pollution is the brightening of the nighttime sky caused by artificial lights.

2. The full moon is about _____ times brighter than first quarter due to the way the sunlight is reflected.

3. Even the slight _____ differences in the air cause motion that can be seen in a telescope, causing ripples to blur the image of any object you are viewing.

4. _____ elevations are generally better than _____ elevations; you're looking through less atmosphere.

5. Much of the "bad seeing" occurs within the telescope itself, so make sure it reaches _____ equilibrium before starting any serious viewing.

6. When observing the skies, it is suggested that you observe the _____ objects first, then the _____ objects later on.

7. For almost all telescopic objects, it is recommended by the author to use the _____ magnification possible.

Short Answers

1. The higher the magnification, the _____ the target will appear, and the _____ the field of view.

2. "_____" is the process of aligning the mirrors of a reflecting telescope.

3. Use _____ vision when looking at faint objects, not looking directly at them.

4. A moment of "good seeing" can occur when objects viewed through a telescope become _____ and exceptionally _____.

5. The best way to find deep sky objects is with a technique called "_____ _____."

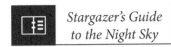

Short Answers

1. The moon looks best in a telescope when it is in or near _____ quarter or _____ quarter.

2. The dividing arc between day and night on the moon is called the "_____."

3. It is common for larger craters on the moon to have a _____ peak.

4. Faults are best viewed near the terminator where their shadows are _____.

5. _____ are large dark regions on the moon.

6. _____ are very light-colored streaks that radiate away from some craters.

7. The "_____ _____" makes the moon appear larger when it is near the horizon.

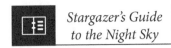
Short Answers

1. Do not use a "_____" filter that is designed for the eyepiece, as these are unsafe.

2. The center of the sun is noticeably _____ than the limb, or outer section.

3. The sun is gaseous and hotter in the _____ than the _____.

4. Sunspots are _____ regions on the sun caused by intense magnetic fields.

5. The temperature of the sun's surface is around _____ degrees Celsius.

6. Sunspots have an ____-year cycle.

7. The sun's _____ magnetic field lies below the surface of the sun, and is oriented in an east-west direction.

Short Answers

1. Jupiter takes about one year on average to move from one _____ to the next.

2. Planets lie roughly in the _____, which marks the path that the sun appears to travel.

3. _____ is a noticeably red object in the sky.

4. Only Venus, the moon, and the sun shine brighter than _____.

5. The best time to see Jupiter is when the planet is in _____, which occurs when the earth is in between Jupiter and the sun.

6. The zones on Jupiter consist of higher altitude clouds made primarily of _____.

7. The four most visible moons of Jupiter are called the _____ satellites.

Short Answers

1. The red spot _____ time is the time the red spot on Jupiter will be centered best for viewing.

2. Jupiter's rapid _____, less than ten hours, is faster than any other planet.

3. Since its moons orbit around its equator, it is very common for the moons to _____ Jupiter.

4. Jupiter has a nearly ___-year orbital period.

5. Venus is only visible in the _____ shortly after sunset or in the _____ shortly before sunrise.

6. Venus is often called "the evening star" or "the morning star" since it dominates the sky when near greatest _____.

7. _____ of Venus happen only twice per century, and the two events are always separated by eight years.

Short Answers

1. Saturn is physically about _____ times the diameter of Earth.

2. Though Saturn is about one _____ miles away from Earth, its distance doesn't change very much.

3. The rings of Saturn comprise trillions of tiny _____ that orbit its equator.

4. The division or "gap" between the A and B rings of Saturn is called the "_____ _____."

5. Saturn has over ____ moons, the largest being _____.

6. The true tilt of Saturn's axis of rotation remains constant at _____ degrees.

Short Answers

1. Saturn takes _____ years to orbit the sun.

2. One can observe a "ring-less" Saturn when the planet is _____-_____.

3. Mars only looks good when it is very near _____.

4. Mars can appear over _____ times larger at opposition.

5. To the unaided eye, Mars appears a vivid _____ color.

6. The _____ Crater was the landing site for the science lab Curiosity.

Short Answers

1. The darkest feature on Mars, _____ _____, passes through Mars' equator.

2. On Mars, the valley _____ often fills up with fog and so appears almost white.

3. The ____ _____ change size with the Martian seasons.

4. The rotation period of Mars is _____ hours and 37 minutes.

5. The _____ _____ is called the Grand Canyon of Mars.

6. Mars has two tiny moons: _____ and _____.

7. Mercury is the closest planet to the sun and takes only _____ days to complete one orbit.

Short Answers

1. Uranus appears as a featureless _____ sphere, lying at a distance of around 1.8 billion miles from Earth.

2. A moderately large backyard telescope can theoretically view five moons of Uranus: Miranda, Ariel, Umbriel, _____, and _____.

3. Neptune is essentially the twin of _____ in physical size and color.

4. The only moon of Neptune visible with a moderately sized backyard telescope is _____.

5. Pluto is now a part of a new class of objects: the Trans-_____ Objects (TNOs).

6. Near _____, Pluto is actually slightly closer to the sun than Neptune.

7. Pluto takes ____ years to orbit the sun once.

Short Answers

1. The brightness of stars as we see them from Earth are classified according to their "_____ _____."

2. In star classification, the lower numbers represent the _____ stars.

3. The _____ _____ of a star is the apparent magnitude that a star would have if it were placed 10 parsecs away from the observer.

4. Stars are classified based on their _____ features, which is the presence or absence of certain wavelengths of light.

5. _____ stars are the hottest, and _____ stars are the coolest.

6. Stars are organized into one of seven classes in order from hottest to coolest: ___, ___, ___, ___, ___, ___, and ___.

7. The seven classes of stars are subdivided into a range of ___ categories within each class (represented by a letter), and based on their _____ within a given temperature class.

Short Answers

1. Double stars are called _____ stars.

2. _____ doubles are two stars that appear to be close together, but are in fact at different distances.

3. _____ binaries are true binaries, two stars that can be separated in a telescope.

4. _____, the southernmost star in the Southern Cross, is a visual binary.

5. The _____-Russell diagram is a graph that plots the absolute magnitude of a star against its surface temperature.

6. _____ mass stars are cool and red, _____ mass stars are yellow, and _____ mass stars are hot and blue.

Short Answers

1. Some of the best galaxies, star clusters, and nebulae that can be viewed by a small telescope are found in the _____ list, first published in 1781.

2. There are two types of star clusters, which are denser regions of stars: _____ clusters and _____ clusters.

3. _____ clusters typically consist of about 100,000 stars and are always spherical in shape.

4. To the unaided eye, the _____ look like a tiny little dipper consisting of about six or seven stars.

5. A number of Messier objects are found along the Milky Way in and around the constellations _____ and _____.

Short Answers

1. M3 is a compact globular star cluster found to the northwest of _____.

2. One of the best globular clusters visible in the Northern Hemisphere, _____, is found in the constellation Hercules.

3. The most spectacular globular cluster visible from Earth is _____ _____.

4. A _____ is a cloud of hydrogen and helium gas spread over a vast region of space.

5. A nebula will glow if _____ are close to it, heating it.

6. A _____ nebula is large and does not have a distinct boundary.

7. A _____ nebula is produced by the ejected gas of a star and is generally smaller than other types of nebulae.

Short Answers

1. The Ring Nebula (M57) is located between Sulafat and Sheliak in the small constellation _____.

2. M27 is called the _____ Nebula because of its quasi-circular, two-lobed structure.

3. The _____ Nebula is a good wintertime planetary nebula, found in the constellation Gemini.

4. The _____ Nebula, M17, is sometimes called the Omega Nebula or Horseshoe Nebula.

5. The Eagle Nebula is well known even among non-experts because of a Hubble image sometimes called "the Pillars of _____."

6. In winter, Hubble's variable nebula (NGC 2261) resembles a _____.

7. A _____ is a collection of around 100 billion stars.

Short Answers

1. The first basic classification of galaxies is the _____ galaxies, which are disk-shaped, with a brighter bulge at the center of the disk.

2. The second basic classification of galaxies, and the one most fall under because of their basic shape, is the _____ galaxies.

3. The third basic classification of galaxies is the _____ galaxies, which are a blend between an elliptical and a spiral.

4. The fourth and final basic classification of galaxies is the _____ galaxies because they do not fit the other specified shapes.

5. Perhaps the best example of a face-on grand design spiral galaxy is the _____ Galaxy (M51).

Short Answers

1. The _____ cluster, more than ten degrees across, presents a view of a large number of galaxies simultaneously.

2. The largest and brightest member of the Virgo cluster is _____.

3. The _____ Galaxy (M104) has a striking dust lane around its perimeter.

4. The worst time to look for galaxies is late summer because of the "zone of _____."

5. _____ are small, blue objects that have enormous redshifts.

6. An _____ _____ _____ (AGN) is a galaxy whose center is unusually bright, and sometimes gives off powerful radio signals.

7. A QSO is a "Quasi-_____ Object."

Short Answers

1. Placing a digital camera up to the eyepiece of a telescope to take a picture is called "_____ imaging."

2. _____ is when the field of view will be reduced in the camera, with soft-edges that are reduced in brightness relative to the center.

3. The best way to do astrophotography is using a CCD, which stands for "_____ _____ device."

4. You generally need three color filters for color photos, which are _____, _____, and _____.

5. For color photos you can also use an SSC, a "_____ _____ color" CCD.

6. A _____ mirror allows you to quickly check on the target object with your eye, and then quickly switch back to CCD imaging mode.

7. A dark frame image is dependent on (1) _____ and (2) time exposure.

Short Answers

1. The higher the temperature, the greater the _____ current will be.

2. In "_____ fielding," if the gain for each photoreceptor is known, you can divide the final image by this value to compensate for the unequal gain of the various pixels.

3. Image _____ is the best way to get long time exposure images on telescopes that have only mediocre tracking.

4. When the photoreceptors are not sensitive to light at all, these are said to be "_____ pixels."

5. For best results in capturing a lunar eclipse, keep the entire moon _____.

6. Star _____ are the tracks, often circular, that appear because of the combination of the earth's rotation and long exposure setting.

Short Answers

1. Only the biblical God as described in the pages of Scripture can make sense of the _____ _____ of nature, which describe the predictable behavior of the cosmos.

2. The fact that natural laws can be expressed by _____ laws confirms that the universe was designed to be understood by the human mind.

3. _____ _____ said that doing astronomy is like "thinking God's thoughts after Him."

4. Did this course meet or exceed your initial expectations? Why or why not?

Astronomy Quizzes

for Use with

The Stargazer's Guide to the Night Sky

Answer Questions: (5 Points Each Question)

1. What is this trend of the earth's rotation called?

2. Name the concept that is useful for understanding the positions and motions of stars.

3. What are constellations called that are close enough to the celestial pole that they are visible all night, year-round?

4. Since stars rise two hours earlier every month, they rise ____ hours earlier after six months.

5. The moon rises (on average) about _____ _____ later each day.

6. The motion of the planets is complicated because their apparent motion in the sky is the combination of their _____ _____ around the sun, plus the _____ _____ in position due to Earth's motion around the sun.

7. Equatorial coordinates are based on the _____ _____. In particular, they are based on the celestial _____.

8. A star wheel or planisphere helps you find stars by lining up the _____ with the _____ on the planisphere.

9. What does the term "equinox" mean?

10. What does the term "solstice" mean?

11. In the Northern Hemisphere, days are longer in the _____ and _____ seasons, and shorter in the _____ and _____ seasons.

12. The configuration of the sun, moon, and Earth is about the same every 18.031 years, and is called the _____ cycle.

13. The earth actually has two shadows — a darker inner shadow called the "_____" and a lighter outer shadow called the "_____."

14. When the moon is farther from the earth, it appears smaller than the sun, and when it passes directly in front of the sun it leaves a thin "ring" or "_____."

15. Planets orbit the sun in slightly _____ paths with the sun at one focus of the ellipse.

16. _____ is when a planet is "behind" the sun.

17. Meteor showers are generally caused by debris left behind by a _____.

18. The most impressive, reliable meteor shower is the _____ meteor shower, occurring around August 12th each year.

19. The sun traces out a thin figure-eight shape across the sky with each day, which is called _____.

20. The constellation in which the sun is found at equinox shifts with time. This phenomenon is called the "_____ _____ _____ _____."

Answer Questions: (10 Points Each Question)

1. The human eye has two different types of light-detecting cells: _____ and _____.

2. When looking at faint objects through a telescope, you must avoid looking _____ at the object.

3. Using a _____ _____ at night helps you see without impacting your dark-adapted vision.

4. As a general guideline, it takes about ____ minutes to dark adapt your eyes.

5. Stars are named based on their _____ and the _____ in which they are found.

6. The constellation Orion is mentioned in the biblical book of _____, dating back to about 2000 B.C.

7. Because its light fades and grows, Algol is often called the "_____ Star."

8. _____ is the brightest star visible in Earth's nighttime sky.

9. Polaris or the "_____ _____" is the one star that doesn't move noticeably with the time of night or time of year.

10. The constellation Corona Australis is also known as the _____ _____.

Answer Questions: (5 Points Each Question)

1. A _____ is when one celestial body passes closely by another.

2. An _____ is when one celestial object passes directly in front of another, blocking the background object from sight.

3. The term "_____" refers to the bright streak in the sky. The rock causing the streak is called the "_____." The rock that hits the earth's surface is called a "_____."

4. During a meteor shower, meteors will appear to move away from the same point in space. This point is called the "_____."

5. A "_____ _____" is a rare occurrence where the sky appears to be raining meteors.

6. The only meteor shower known to be associated with an asteroid rather than a comet is the _____.

7. Approximately every 33 years, the _____ produce a meteor storm.

8. The reason rainbows are colorful is because the water droplets split the sunlight into its constituent _____.

9. A _____ _____ is caused by the reflection and refraction of sunlight off of ice crystals in thin, high cirrus clouds.

10. The word *corona* means "_____."

11. _____ orbit the earth in about 90 minutes.

12. The iridium _____ is the temporary brightening of a satellite associated with the company Iridium Communications Inc.

13. The overall ability of a telescope is determined by the diameter of the primary lens or primary mirror — the "_____."

14. The two kinds of telescopes are _____ and reflectors.

15. "_____ aberration" is when bright objects you view through a refractor are surrounded by a little purple halo.

16. Reflectors use a curved _____ instead of a primary lens to bring light to a focus.

17. Binoculars or a small telescope will allow you to observe the moon, bright stars, open star clusters, the _____ Galaxy, and some bright comets.

18. A _____-_____ is a motor that slowly rotates the telescope in the opposite direction that Earth rotates.

19. Telescope mounts come in two varieties: _____ mounts and _____ (alt-azimuth) mounts.

20. A _____ adds a laser-produced artificial spot, concentric circles, or crosshairs that indicate exactly where the telescope is pointed.

Answer Questions: (10 Points Each Question)

1. The clearest nights will generally be the _____ nights.

2. The full moon is about _____ times brighter than first quarter due to the way the sunlight is reflected.

3. Much of the "bad seeing" occurs within the telescope itself, so make sure it reaches _____ equilibrium before starting any serious viewing.

4. For almost all telescopic objects, it is recommended by the author to use the _____ magnification possible.

5. The higher the magnification, the _____ the target will appear, and the _____ the field of view.

6. "_____" is the process of aligning the mirrors of a reflecting telescope.

7. Use _____ vision when looking at faint objects, not looking directly at them.

8. The best way to find deep sky objects is with a technique called "_____ _____."

9. The dividing arc between day and night on the moon is called the "_____."

10. _____ are large dark regions on the moon.

Answer Questions: (4 Points Each Question)

1. Planets lie roughly in the _____, which marks the path that the sun appears to travel.

2. Only Venus, the moon, and the sun shine brighter than _____.

3. The zones on Jupiter consist of higher altitude clouds made primarily of _____.

4. The four most visible moons of Jupiter are called the _____ satellites.

5. The red spot _____ time is the time the red spot on Jupiter will be centered best for viewing.

6. Jupiter has a nearly _____-year orbital period.

7. Venus is often called "the evening star" or "the morning star" since it dominates the sky when near greatest _____.

8. _____ of Venus happen only twice per century, and the two events are always separated by eight years.

9. Saturn is physically about _____ times the diameter of Earth.

10. The division or "gap" between the A and B rings of Saturn is called the "_____ _____."

11. One can observe a "ring-less" Saturn when the planet is _____-_____.

12. Mars only looks good when it is very near _____.

13. The _____ Crater was the landing site for the science lab Curiosity.

14. On Mars, the valley _____ often fills up with fog and so appears almost white.

15. Mars has two tiny moons: _____ and _____.

16. Mercury is the closest planet to the sun and takes only _____ days to complete one orbit.

17. A moderately large backyard telescope can theoretically view five moons of Uranus: Miranda, Ariel, Umbriel, _____, and _____.

18. The only moon of Neptune visible with a moderately sized backyard telescope is _____.

19. Pluto is now a part of a new class of objects: the Trans-_____ Objects (TNOs).

20. Near _____, Pluto is actually slightly closer to the sun than Neptune.

21. The brightness of stars as we see them from Earth are classified according to their "_____ _____."

22. The seven classes of stars are subdivided into a range of _____ categories within each class (represented by a letter), and based on their _____ within a given temperature class.

23. Double stars are called _____ stars.

24. _____, the southernmost star in the Southern Cross, is a visual binary.

25. The _____-Russell diagram is a graph that plots the absolute magnitude of a star against its surface temperature.

Answer Questions: (10 Points Each Question)

1. Some of the best galaxies, star clusters, and nebulae that can be viewed by a small telescope are found in the _____ list, first published in 1781.

2. There are two types of star clusters, which are denser regions of stars: _____ clusters and _____ clusters.

3. A _____ is a cloud of hydrogen and helium gas spread over a vast region of space.

4. The Eagle Nebula is well known even among non-experts because of a Hubble image sometimes called "the Pillars of _____."

5. The first basic classification of galaxies is the _____ galaxies, which are disk-shaped, with a brighter bulge at the center of the disk.

6. The second basic classification of galaxies, and the one most fall under because of their basic shape, is the _____ galaxies.

7. Star _____ are the tracks, often circular, that appear because of the combination of the earth's rotation and long exposure setting.

8. Only the biblical God as described in the pages of Scripture can make sense of the _____ _____ of nature, which describe the predictable behavior of the cosmos.

9. The fact that natural laws can be expressed by _____ laws confirms that the universe was designed to be understood by the human mind.

10. _____ _____ said that doing astronomy is like "thinking God's thoughts after Him."

Answer Keys

for Use with

The Stargazer's Guide to the Night Sky

The Stargazer's Guide to the Night Sky ⚷ Worksheet Answer Keys

Introduction – Worksheet 1

1. Answers will vary.
2. With your eyes alone, with binoculars, or with a telescope.
3. Yes. There are many similarities, but star charts can vary.
4. Yes. The book includes star charts depending on the seasonal skies.
5. The moon and the sun.

Chapter 1 – Worksheet 1

1. This is because of the earth's rotation on its axis; because the earth is spinning in the opposite direction.
2. It is called "diurnal motion."
3. It is an approximate 24-hour cycle.
4. It is called the "celestial sphere."
5. It is called the "celestial equator."
6. They are called "circumpolar" constellations.
7. Sidereal day

Chapter 1 – Worksheet 2

1. The sidereal day is the true rotation rate of Earth as seen from a distant star. The solar day is how long it takes for the sun to return to its highest point in the sky as viewed from a location on Earth.
2. 12
3. It washes out nearly everything else in the sky.
4. 50 minutes
5. The gravitational pull of the sun on the moon is about twice the pull of the earth on the moon.
6. percentage
7. 29.3, 27.5

Chapter 1 – Worksheet 3

1. true motion, apparent shift
2. The earth's rotation axis that is tilted relative to its orbit around the sun by 23.4 degrees.

3. The first is based on our local horizon. The second is based on the celestial sphere.
4. Altitude
5. Azimuth
6. celestial sphere, equator

Chapter 1 – Worksheet 4

1. south, north
2. day, year
3. setting
4. It's best to get outside and watch.
5. date, time
6. planets, moon

Chapter 2 – Worksheet 1

1. declination, 12
2. It means "equal night."
3. It means "sun stop."
4. Arctic
5. Spring/summer, fall/winter
6. Ecliptic

Chapter 2 – Worksheet 2

1. On the spring equinox
2. Solar, lunar
3. Node
4. Saros
5. Umbra, penumbra
6. Photosphere
7. 400

Chapter 2 – Worksheet 3

1. Annulus
2. Libration
3. Elliptical
4. Superior
5. Conjunction

6. Counterclockwise

7. Mercury, Venus

Chapter 2 – Worksheet 4

1. Solar

2. Elliptical

3. Comet

4. Perseid

5. Analemma

6. Precession of the equinoxes

Chapter 3 – Worksheet 1

1. Rods, cones

2. Light

3. Directly

4. Rhodopsin

5. Red light

6. 30

7. Carrots

Chapter 4 – Worksheet 1

1. Brightness, constellation

2. Job

3. Asterisms

4. Hottest, brightest

5. Demon

6. Sirius

7. Dog Star

Chapter 4 – Worksheet 2

1. Leo

2. North Star

3. Mizar

4. Thuban

5. Scorpius

Chapter 4 – Worksheet 3

1. Triangle

2. Deneb

3. Southern Crown

4. Square

5. bullhorn

6. M31

Chapter 5 – Worksheet 1

1. Conjunction

2. Massing

3. Triple conjunction

4. Occultation

5. Lunar

6. Lunar, solar

Chapter 5 – Worksheet 2

1. Projection

2. Heated

3. Meteor, meteoroid, meteorite

4. Radiant

5. Fireball

6. Train

7. Meteor storm

Chapter 5 – Worksheet 3

1. Geminids

2. Quadrantids

3. Eta Aquarids, Orionids

4. Leonids

5. Wavelengths

6. Solar halo

Chapter 5 – Worksheet 4

1. Crown

2. Zodiacal light, gegenchein

3. Satellites

4. Winter

5. Flare

Chapter 6 – Worksheet 1

1. Bright
2. Resolving
3. Aperture
4. Refractors
5. Reflectors
6. Refractors
7. Chromatic

Chapter 6 – Worksheet 2

1. Mirror
2. Isaac Newton
3. Schmidt-Cassegrain
4. Dewing
5. Dew shield

Chapter 6 – Worksheet 3

1. Andromeda
2. Mount
3. Clock-drive
4. Equatorial, horizon
5. Spotting (or "spotter")
6. Telrad
7. Computer

Chapter 6 – Worksheet 4

1. During the day
2. North Star
3. Auto alignment
4. Invert, magnify, brighten

Chapter 7 – Worksheet 1

1. Coldest
2. Red
3. Green
4. Convection
5. Half hour
6. Dew shield

Chapter 7 – Worksheet 2

1. Light
2. Nine
3. Temperature
4. Higher, lower
5. Thermal
6. Brighter, fainter
7. Lowest

Chapter 7 – Worksheet 3

1. Blurrier, smaller
2. Collimation
3. Averted
4. Still, clear
5. Star hopping

Chapter 8 – Worksheet 1

1. First, third
2. Terminator
3. Central
4. Longest
5. Maria
6. Ejecta
7. Moon illusion

Chapter 8 – Worksheet 2

1. Solar
2. Brighter
3. Interior, exterior
4. Darker
5. 6,000
6. 1
7. Toroidal

Chapter 9 – Worksheet 1

1. Constellation
2. Ecliptic
3. Mars
4. Jupiter

5. Opposition

6. Ammonia

7. Galilean

Chapter 9 – Worksheet 2

1. Transit

2. Rotation

3. Eclipse

4. 12

5. West, east

6. Elongation

7. Transits

Chapter 9 – Worksheet 3

1. Nine

2. Billion

3. Moonlets

4. Cassini division

5. 60, Titan

6. 26.7

Chapter 9 – Worksheet 4

1. 29.5

2. Edge-on

3. Opposition

4. Five

5. Red

6. Gale

Chapter 9 – Worksheet 5

1. Syrtis Major

2. Hellas

3. Ice caps

4. 24

5. Valles Marineris

6. Phobos, Deimos

7. 88

Chapter 9 – Worksheet 6

1. Blue

2. Titania, Oberon

3. Uranus

4. Triton

5. Neptunian

6. Perihelion

7. 248

Chapter 10 – Worksheet 1

1. Apparent magnitude

2. Brighter

3. Absolute magnitude

4. Spectral

5. Blue, red

6. O, B, A, F, G, K, and M

7. Ten, size

Chapter 10 – Worksheet 2

1. Binary

2. Optical

3. Visual

4. Acrux

5. Hertzsprung

6. Low, medium, high

Chapter 11 – Worksheet 1

1. Messier

2. Open, globular

3. Globular

4. Pleiades

5. Sagittarius, Scorpius

Chapter 11 – Worksheet 2

1. Arcturus

2. M13

3. Omega Centauri

4. Nebula

5. Stars

6. Diffuse

7. Planetary

Chapter 11 – Worksheet 3

1. Lyra

2. Dumbbell

3. Eskimo

4. Swan

5. Creation

6. Comet

7. Galaxy

Chapter 11 – Worksheet 4

1. Spiral

2. Elliptical

3. Lenticular

4. Irregular

5. Whirlpool

Chapter 11 – Worksheet 5

1. Virgo

2. M87

3. Sombrero

4. Avoidance

5. Quasars

6. Active galactic nucleus

7. Stellar

Chapter 12 – Worksheet 1

1. Afocal

2. Vignetting

3. Charge coupled

4. Red, green, blue

5. Single shot

6. Flip

7. Temperature

Chapter 12 – Worksheet 2

1. Dark

2. Flat

3. Stacking

4. Dead

5. Framed

6. Trails

Chapter 12 – Worksheet 3

1. Universal laws

2. Mathematical

3. Johannes Kepler

4. Answers will vary.

Quiz 1 – Chapters 1–2

1. It is called "diurnal motion."
2. It is called the "celestial sphere."
3. They are called "circumpolar" constellations.
4. 12
5. 50 minutes
6. True motion, apparent shift
7. Celestial sphere, equator
8. Date, time
9. It means "equal night."
10. It means "sun stop."
11. Spring/summer, fall/winter
12. Saros
13. Umbra, penumbra
14. Annulus
15. Elliptical
16. Conjunction
17. Comet
18. Perseid
19. Analemma
20. Precession of the equinoxes

Quiz 2 – Chapters 3–4

1. Rods, cones
2. Directly
3. Red light
4. 30
5. Brightness, constellation
6. Job
7. Demon
8. Sirius
9. North Star
10. Southern Crown

Quiz 3 – Chapters 5–6

1. Conjunction
2. Occultation

3. Meteor, meteoroid, meteorite
4. Radiant
5. Meteor storm
6. Geminids
7. Leonids
8. Wavelengths
9. Solar halo
10. Crown
11. Satellites
12. Flare
13. Aperture
14. Refractors
15. Chromatic
16. Mirror
17. Andromeda
18. Clock-drive
19. Equatorial, horizon
20. Telrad

Quiz 4 – Chapters 7–8

1. Coldest
2. Nine
3. Thermal
4. Lowest
5. Blurrier, smaller
6. Collimation
7. Averted
8. Star hopping
9. Terminator
10. Maria

Quiz 5 – Chapters 9–10

1. Ecliptic
2. Jupiter
3. Ammonia
4. Galilean
5. Transit

6. 12
7. Elongation
8. Transits
9. Nine
10. Cassini division
11. Edge-on
12. Opposition
13. Gale
14. Hellas
15. Phobos, Deimos
16. 88
17. Titania, Oberon
18. Triton
19. Neptunian
20. Perihelion
21. Apparent magnitude
22. Ten, size
23. Binary
24. Acrux
25. Hertzsprung

Quiz 6 – Chapters 11–12

1. Messier
2. Open, globular
3. Nebula
4. Creation
5. Spiral
6. Elliptical
7. Trails
8. Universal laws
9. Mathematical
10. Johannes Kepler